Cambridge Primary

Hodder Cambridge Primary
Maths
Activity Book
A

Foundation Stage

Ann and Paul Broadbent

HODDER
EDUCATION
AN HACHETTE UK COMPANY

Orders: please contact Hachette UK Distribution, Hely Hutchinson Centre, Milton Road, Didcot, Oxfordshire, OX11 7HH. Telephone: +44 (0)1235 827827. Email education@hachette.co.uk Lines are open from 9 a.m. to 5 p.m., Monday to Friday. You can also order through our website: www.hoddereducation.com

© Ann Broadbent and Paul Broadbent 2018

First published in 2018

This edition published in 2018 by Hodder Education,
An Hachette UK Company
Carmelite House
50 Victoria Embankment
London EC4Y 0DZ
www.hoddereducation.co.uk

Impression number   10 9 8 7 6 5 4 3

Year   2022   2021

Cover illustration by Steve Evans

Illustrations by Jeanne du Plessis, Vian Oelofsen

Typeset in FS Albert 17 pt by Lizette Watkiss

Printed in the United Kingdom

A catalogue record for this title is available from the British Library.

ISBN 978 1 5104 3182 9

MIX
Paper from
responsible sources
FSC™ C104740

Contents

Sorting

⭐ Sort these clothes by type or pattern. Join them to a matching box.

⭐ Find and colour matching keys. Choose a different colour for each type.

⭐ Which things go together? Sort these in different ways.

Matching

⭐ Which socks go together? Colour to match each pair of socks.

⭐ These shoes are mixed up. Draw lines to join the matching pairs.

Counting how many

⭐ Count how many there are of each animal.

⭐ Count the petals on each flower.
Colour to show the flowers that are the same.

Counting to 5

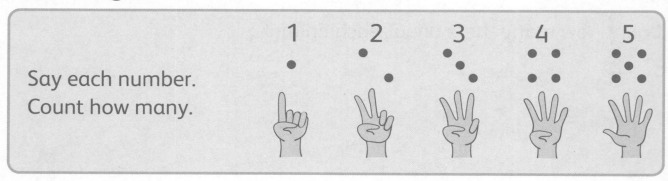

Say each number.
Count how many.

⭐ Count these things. Write the numbers.

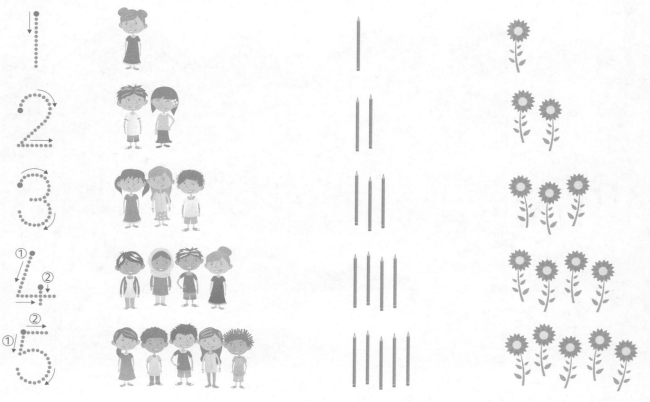

⭐ How many birds are on each branch?

Matching numbers

⭐ Count the carriages. Write the last number in each count.

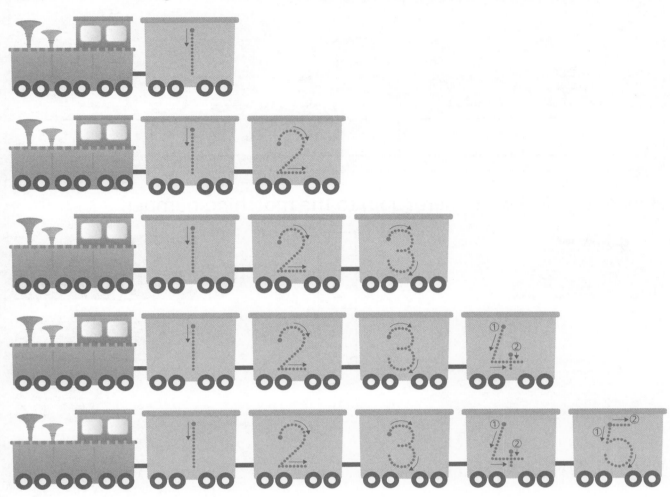

⭐ Colour some of these cars.
How many have you coloured? Write the number to match.

Counting and numbers to 5

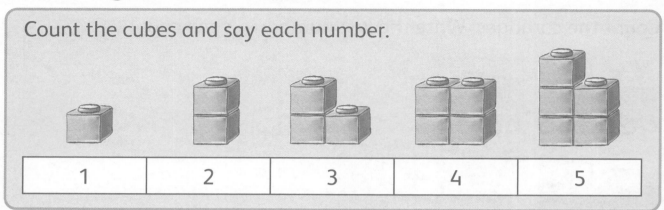

Count the cubes and say each number.

| 1 | 2 | 3 | 4 | 5 |

⭐ How many bricks are there? Join to the matching number.

| 1 |
| 2 |
| 3 |
| 4 |
| 5 |

⭐ Write the numbers.

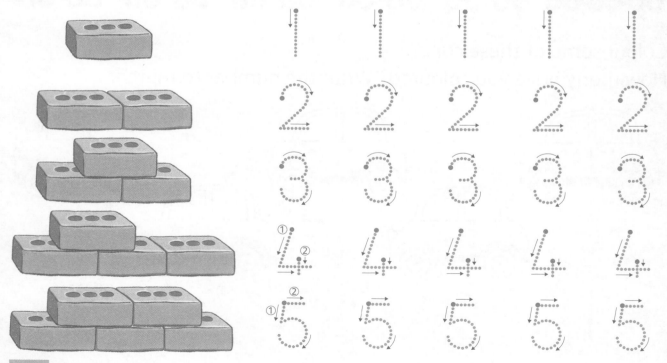

Read these numbers. Join each child to the matching number of balloons.

 Write these numbers. Draw candles on each cake to match.

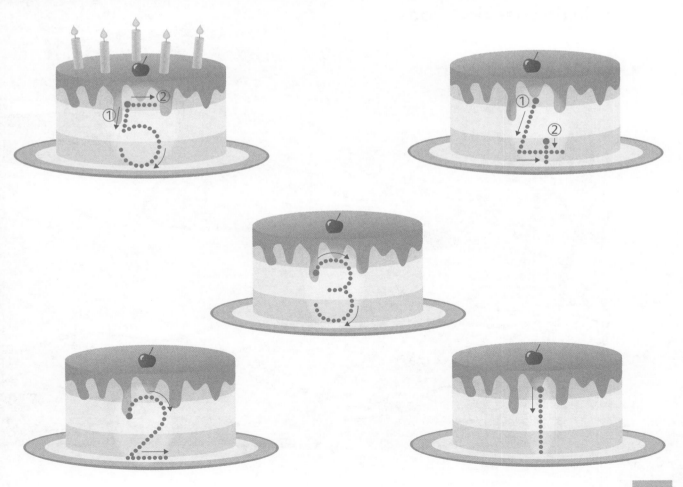

Counting on and back

Read this rhyme together. Repeat the rhyme until there are no frogs left on the speckled log.

Five little speckled frogs
Sat on a speckled log,
Eating the most delicious bugs – yum, yum!
One jumped into the pool,
Where it was nice and cool,
Then there were four speckled frogs – glug, glug!

Four little speckled frogs ...

Three little speckled frogs ...

Two little speckled frogs ...

One little speckled frog ...

⭐ How many of these can you count in the picture?
Circle the number to match.

🦆	1	2	3	4	5
🪷	1	2	3	4	5
🦟	1	2	3	4	5
🐟	1	2	3	4	5
🐸	1	2	3	4	5

Zero to five

These all show zero.

0

⭐ Find the empty bowls. Join them to 0.

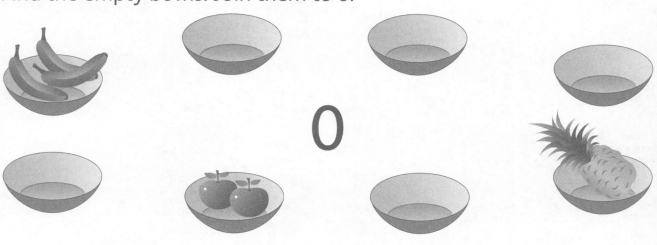

0

⭐ Join numbers that are the same.

⭐ Put counters in these five frames to show each number.
What do you notice?

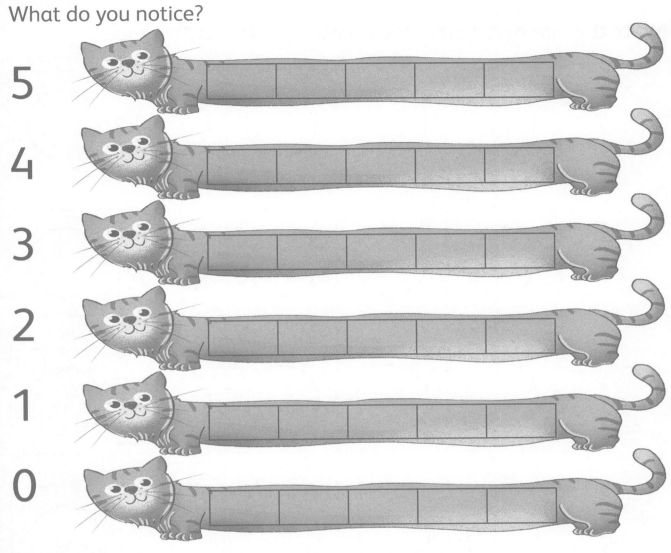

5

4

3

2

1

0

⭐ Colour the frames on each boat to show its number.

3

5

2

15

Writing numbers

⭐ Write the numbers. Colour the circles to match each number.

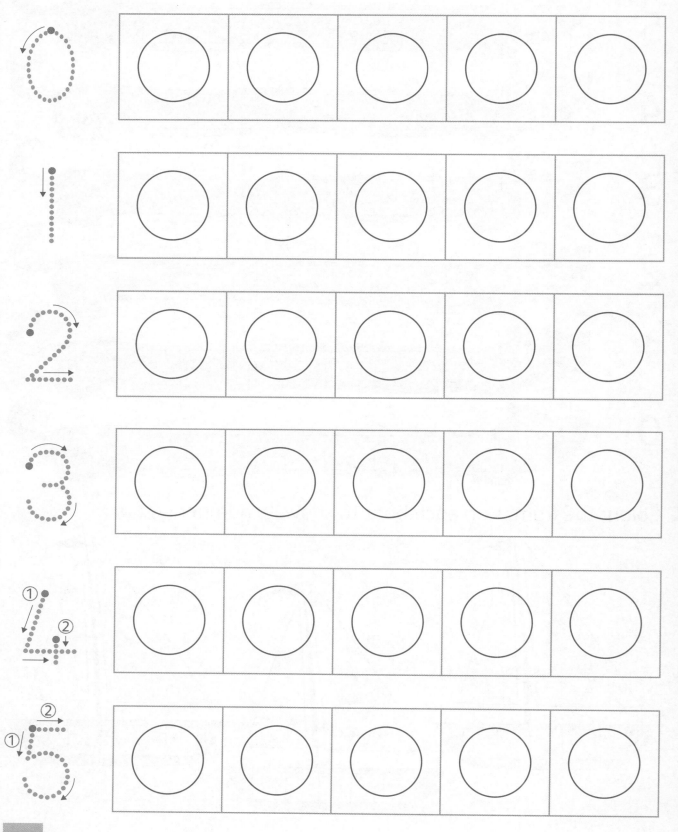

⭐ How many in each bowl? Count and write the number.

⭐ Draw some fruit in this bowl. Write the number.

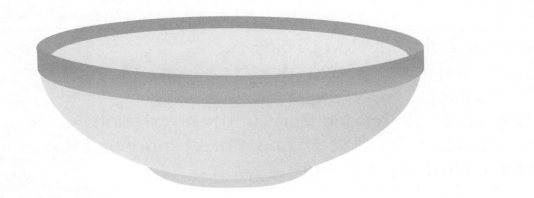

One more and one less

⭐ Play the Bus Stop Game!

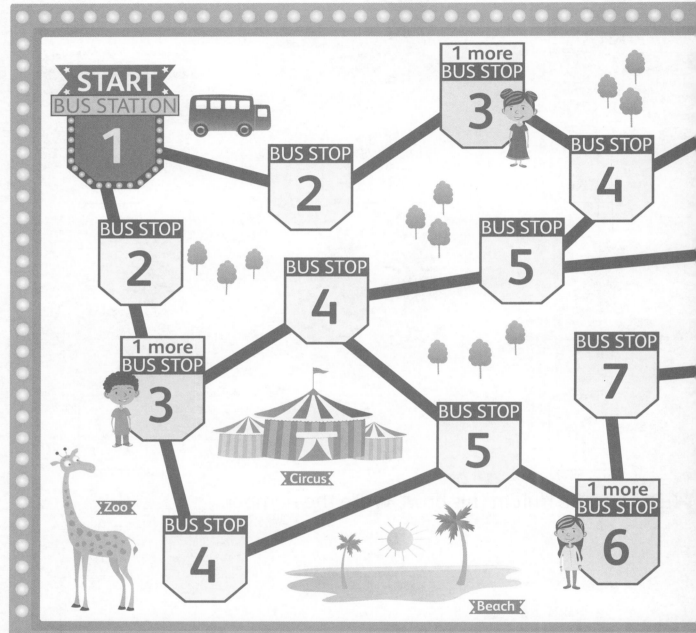

This is a game for 2 players. How to play:

- Write **1 more** and **1 less** on two stickers and stick them onto each side of a coin.
- Each player has a counter and starts at the bus station.
- Players take turns to flip the coin and move **1 more** or **1 less**.
- A player must flip **1 more** to start.

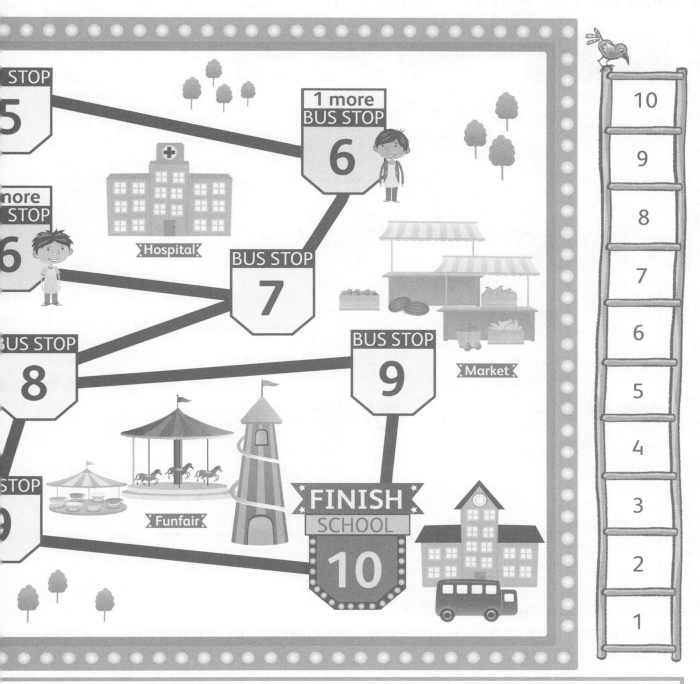

- Players should move from the bus station in different directions.
- Say the numbers as you move, such as **4 and 1 more is 5**, or **1 less than 8 is 7**. Use the number ladder to help you.
- If you land on a bus stop with a passenger, add **1 more** and move your counter.
- The winner is the first player to reach the school.

Putting together

⭐ Count each set. Add them together. Write each total.

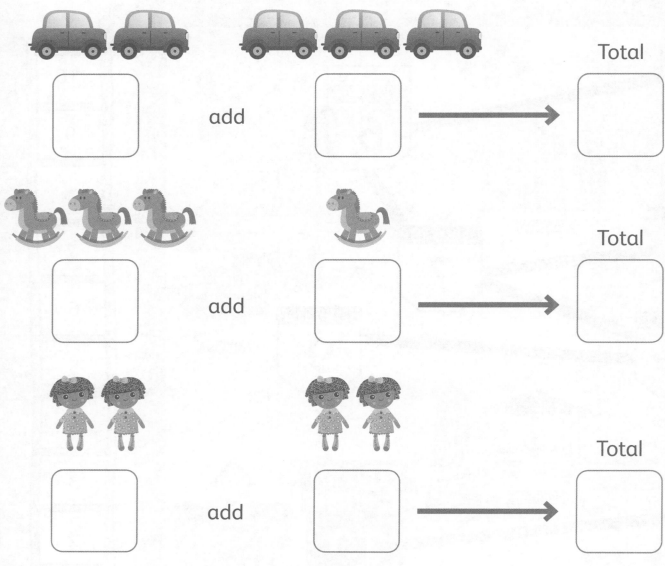

Total

add

Total

add

Total

add

⭐ Draw two more spots on each leaf.
Write how many spots that makes altogether on each leaf.

Making totals

Look how to make 4.

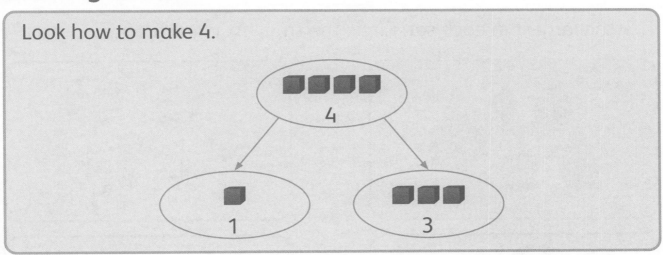

⭐ Make 4 in different ways.

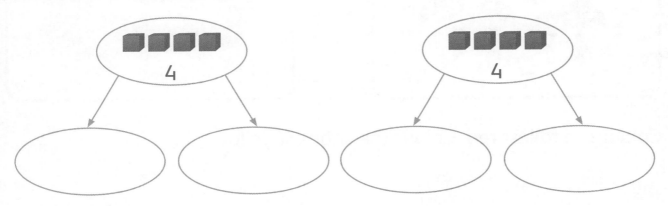

⭐ Use two colours to make 5. Write over the number.

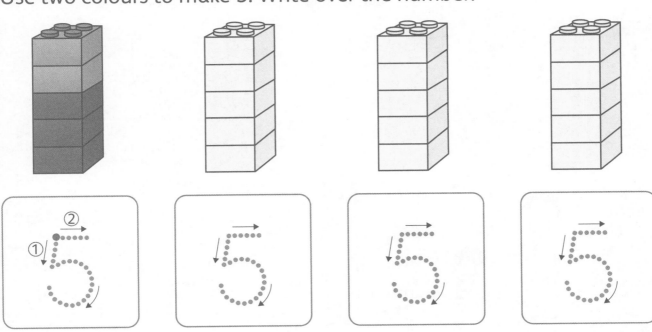

Comparing objects

⭐ Tick the largest in each set. Circle the smallest in each set.

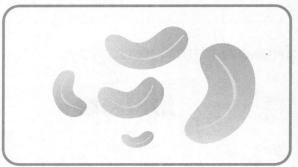

⭐ Colour the tallest red. Colour the shortest yellow.

Comparing length

⭐ Tick the longest fishing line. Circle the shortest fishing line.

⭐ Colour the longest fish above red. Colour the shortest fish above yellow.

⭐ Draw a fish longer than this one.

Long and short

⭐ Join the words to the pictures.

longer   shorter

shorter   taller

shorter   taller

longer   shorter

shorter   longer

taller   shorter

Day and night

⭐ Which of these pictures show day time? Which show night time?

⭐ Do you know the days of the week?

Monday	Tuesday	Wednesday	Thursday	Friday	Saturday	Sunday
1	2	3	4	5	6	7

I spy ... shapes

⭐ Look for these shapes.

circle

rectangle

square

⭐ Join the shapes that are the same.

⭐ Draw these shapes.

| square | triangle | rectangle |

Patterns

 Talk about the patterns in this picture.

Follow these patterns with your pencil.

 Draw your own pattern.

Solid shapes

⭐ Which shapes can you see?

⭐ Find the matching shapes.

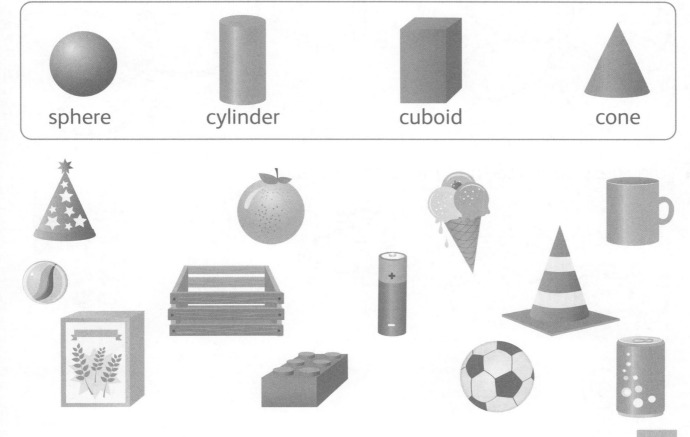

sphere   cylinder   cuboid   cone

What can you remember?

⭐ Which balls are the same? Draw a line to join matching balls.

⭐ Draw spots on each wing. Give each butterfly a total of 5 spots.

⭐ Join these cubes to their matching numbers.

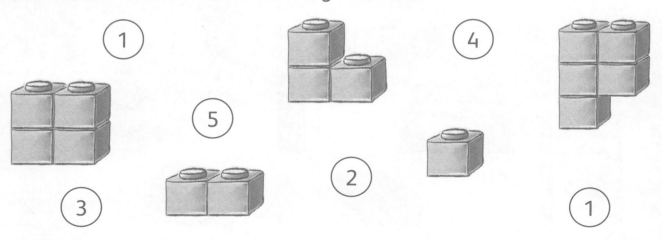

⭐ Draw two more spots on each top.
Count how many spots that makes altogether on each top.

⭐ Draw some long tails on these monkeys.
Which tail is the longest? Which tail is the shortest?

⭐ What is the same about these shapes? What is different?

Self-assessment

Colour the stars to show what you can do!

Sorting and counting	I can sort objects into two groups.	☆
	I can count how many cubes there are in a set.	☆
Numbers to 5	I can count and match objects to a number card.	☆
	I can read the numbers to 5.	☆
Counting forwards and backwards	I can count to 5 and then back again to zero.	☆
	I know that 0 means zero or nothing.	☆
	I can write the numbers to 5.	☆
Addition and subtraction	I can add one more to a set and say how many.	☆
	I can take one away from a set and say what is left.	☆
	I can put two groups of objects together and count the total.	☆
Measures and time	I can compare the length of two objects and say which is longer.	☆
	I know the days of the week.	☆
Shapes and patterns	I can describe what flat shapes look like.	☆
	I can copy and draw different patterns.	☆